Astro Bot;

The Defenders of Space Travel

James Peterson

Copyright

All right reserved. No part of this publication may be reproduced, distributed, or transmitted in any form or by any means, including photocopying, recording, or other electronic or mechanical methods, without the prior written permission of the publisher, except in the case of brief quotations embodied in critical reviews and certain other noncommercial uses permitted by copyright law.

Copyright © James Peterson, 2024.

Table of contents

Chapter 1: Introduction to Astro Bot 6
    What is an Astro Bot? 6

Chapter 2: History of Space Robotics 11
    Early Developments in Space Robots 11
    Evolution of Space Bots: From Probes to Advanced Robotics 13
    Notable Robotic Missions and Their Impact 15

Chapter 3: Designing an Astro Bot 19
    Key Components of an Astro Bot 19
    Sensors and Cameras 22
    AI and Machine Learning Systems 23
    Building a Bot for Space: Challenges and Solutions 25
    Human-Robot Interaction: A New Frontier 26

Chapter 4: Astrobot Abilities and Functions 27
    Autonomous Navigation in Space 28
    Surface Exploration and Sample Collection 29
    Repair and Maintenance of Spacecraft 31
    Communication and Data Relay Capabilities 33

Chapter 6: Mission Planning and Operations 35
    How to Prepare an Astrobot for Space Missions 36
    Deploying Bots on Planetary Surfaces and in Orbit 38

  Real-Time Adaptation and Problem Solving............... 40

  Cooperation with Human Astronauts........................... 42

Chapter 7: Astro Bot Adventures.. 44

 Module 1: Journey to Mars... 44

  Preparing for the Mission 44

  Landing on the Red Planet 45

  Exploring the Martian Landscape 46

 Module 2: Secrets of the Asteroid Belt........................ 47

  Navigating Through Space Rocks 47

  Sampling Unknown Asteroids 48

  Bringing Back the Unknown to Earth 49

 Module 3: The Search for Life on Europa.................... 49

  Entering Jupiter's Domain... 50

  Drilling into the Icy Surface 50

  Detecting Signs of Life 51

  The Future of Astrobots in Space Exploration 52

 Upcoming Missions and the Role of Astrobots............. 52

 Potential for Human-Robot Collaboration in Deep Space 55

Chapter 8: Astrobots in Interstellar Travel and Colonization... 57

 Ethical Considerations and Challenges........................ 59

 Privacy and Security Concerns....................................... 60

The Ethics of Autonomy and Decision-Making 62

The Risk of Malfunction or Rogue Behavior 64

Chapter 9: Conclusion; The Way Forward 68

Summary of Key Takeaways 68

Vision for the Future of Astro Bots 70

Inspiration for the Next Generation of Space Explorers
.... 72

Chapter 1: Introduction to Astro Bot

What is an Astro Bot?

Astro *Bot* is an advanced robotic system specifically designed for space exploration and operations beyond Earth. These bots are equipped with cutting-edge technology, including artificial intelligence (AI), machine learning, autonomous navigation, and specialized tools for analyzing extraterrestrial environments. The term "Astro Bot" can refer to a wide range of robotic devices, from rovers exploring the surface of distant planets to robotic arms assisting astronauts on space stations.

Astrobots are more than just mechanical devices; they are intelligent, adaptable, and resilient systems capable of performing complex tasks in the harsh and unpredictable conditions of space. Unlike traditional robots, Astro Bots are designed to work autonomously or semi-autonomously, handling everything from collecting soil samples on Mars to maintaining and repairing spacecraft in orbit. These robots are at the forefront of space technology, bridging the gap between human capabilities and the vast, unexplored frontiers of space.

The Concept and Vision Behind Astro Bots

The concept of Astro Bots revolves around enhancing our ability to explore and understand space without putting human lives at risk or incurring the enormous costs associated with manned missions. The vision behind Astro Bots is rooted in creating a new generation of space explorers that can go where humans cannot endure conditions that humans cannot survive, and perform tasks that would be too dangerous or tedious for human astronauts.

Astrobots are envisioned as the ideal companions for future space missions, working alongside human astronauts to accomplish complex tasks more efficiently. They can explore unknown terrains, conduct scientific experiments, repair equipment, and build infrastructure for future human settlements.

The ultimate goal is to create a fleet of versatile and intelligent robots that can autonomously adapt to various space environments, learn from their experiences, and make decisions in real time to achieve mission objectives.

By developing Astro Bots, scientists and engineers aim to expand our reach into space, from our solar system to potentially other star systems, without the limitations imposed by human biology. These bots are crucial in humanity's quest to understand the universe, find extraterrestrial life, and establish a human presence beyond Earth.

The Need for Robotics in Space Exploration

The exploration of space is fraught with numerous challenges, many of which make human exploration impractical or impossible shortly. The vast distances, hostile environments, extreme temperatures, high levels of radiation, and the lack of a breathable atmosphere are just a few of the factors that make space one of the most inhospitable frontiers for humans. This is where robotics plays a vital role.

1. **Safety and Risk Management:** Sending robots instead of humans on exploratory missions significantly reduces the risk to human life. Astrobots can endure conditions that would be fatal for humans, such as intense radiation fields, extreme cold or heat, and lack of oxygen. In situations where human presence is too risky, Astro Bots provide a safer alternative.

1. **Cost-Effectiveness:** Manned missions are incredibly expensive, requiring extensive life-support systems, food, water, and safe return capabilities. Robots, on the other hand, do not need these resources, making missions much more cost-effective. Astrobots can be deployed for long durations without the need for rest or sustenance, reducing the logistical complexities associated with human spaceflight.

2. **Extended Reach and Accessibility:** Many areas of our solar system are currently inaccessible to humans due to technological and biological limitations.

Astrobots can explore places like the surface of Venus, the ice-covered oceans of Europa, or the deep craters of the Moon, where extreme conditions prevent human exploration. They can operate for extended periods in environments that would be impossible for humans to endure.

3. **Precision and Reliability:** Astro Bots are capable of performing repetitive, precise tasks with high accuracy. They are designed to conduct scientific experiments, collect samples, and perform analyses that require steady hands and meticulous attention to detail. Their reliability is enhanced by their ability to operate continuously without fatigue, increasing the efficiency of missions.

4. **Support for Human Missions:** As we advance toward human missions to Mars and beyond, Astro Bots will play a critical support role. They can be used to scout landing sites, construct habitats, and set up communication systems before human arrival. During manned missions, they can work alongside astronauts, reducing the

workload and assisting with tasks that require dexterity and precision.

Chapter 2: History of Space Robotics

Early Developments in Space Robots

The history of space robotics began in the mid-20th century when the world's space agencies started to explore the possibilities of sending unmanned machines into space. The earliest space robots were not the advanced machines we envision today but rather simple probes and satellites designed to gather basic data from beyond Earth's atmosphere.

The first step in space robotics came with the launch of Sputnik 1 by the Soviet Union in 1957, which marked the dawn of the space age. Although Sputnik 1 was a simple radio-transmitting satellite, it laid the foundation for more sophisticated robotic missions. This was followed by the Luna series, also launched by the Soviet Union, which achieved several historic milestones, including the first robotic spacecraft to reach the Moon, take pictures of its far side, and soft-land on its surface.

In the 1960s, NASA's Mariner program began to push the boundaries of robotic exploration further. The Mariner 2 mission, launched in 1962, became the first successful flyby of another planet, Venus. Mariner 4, launched in 1964, achieved the first successful flyby of Mars, sending back the first close-up images of the Martian surface. These early missions demonstrated the feasibility of using robots to explore other worlds, setting the stage for more advanced robotic systems.

Robotic exploration continued to evolve with the development of the Surveyor and Viking programs. In 1966, Surveyor 1 successfully landed on the Moon and transmitted thousands of images back to Earth, providing crucial data for future manned missions. In 1976, NASA's Viking

1 and Viking 2 became the first successful landers on Mars, conducting extensive studies of the Martian soil and atmosphere and searching for signs of life. These missions were instrumental in demonstrating that robots could perform complex tasks, including landing, navigation, and scientific analysis, in the harsh environments of space.

Evolution of Space Bots: From Probes to Advanced Robotics

As technology advanced, space robots evolved from simple probes and satellites to sophisticated, multifunctional robotic systems capable of performing a wide range of tasks. The development of more advanced robotic technology was driven by the need to explore more distant and hostile environments in our solar system.

In the late 1970s and early 1980s, the focus shifted from flybys and basic landers to more complex exploration missions. NASA's Voyager 1 and Voyager 2, launched in 1977, were revolutionary robotic spacecraft designed for the exploration of the outer planets. These spacecraft carried a suite of scientific instruments and were designed to perform multiple flybys of Jupiter,

Saturn, Uranus, and Neptune. They provided an unprecedented wealth of data, including detailed images and measurements of these planets and their moons. The Voyager missions showcased the potential for long-duration robotic exploration and paved the way for future missions beyond our solar system.

The 1990s marked a new era of robotic exploration with the advent of mobile robotic platforms, or rovers. NASA's Sojourner, part of the Mars Pathfinder mission in 1997, became the first robotic rover to operate on another planet. Although small and relatively simple by today's standards, Sojourner demonstrated the effectiveness of robotic mobility on the Martian surface, allowing scientists to explore a wider area and conduct detailed geological studies.

Following the success of Sojourner, more advanced rovers were developed, such as NASA's Spirit and Opportunity rovers, which landed on Mars in 2004. These twin rovers were designed to travel long distances and conduct in-depth analyses of Martian rocks and soil. They exceeded all expectations, with Opportunity operating for nearly 15 years, far beyond its planned 90-day mission. These rovers provided critical insights into the planet's geology and history, including evidence of past water activity.

The evolution of space robotics continued with the development of increasingly sophisticated autonomous systems. The Curiosity rover, which landed on Mars in 2012, was equipped with advanced scientific instruments, a robotic arm, and the capability to analyze samples on-site. This marked a significant leap in robotic autonomy and scientific capability. The latest addition to this lineage, Perseverance, landed on Mars in 2021 with even more advanced technologies, including a small helicopter named Ingenuity, the first powered flight on another planet, showcasing the ongoing evolution of space robots from ground-based exploration to aerial reconnaissance.

Notable Robotic Missions and Their Impact

Throughout history, several robotic missions have made significant contributions to our understanding of space and have had profound impacts on science and technology:

1. Luna 2 (1959): The first human-made object to impact the Moon. It marked a major milestone in

space exploration and laid the groundwork for future lunar missions.

2. Mariner 10 (1973): The first spacecraft to use a gravity-assisted maneuver and the first to explore two planets (Venus and Mercury). It provided detailed images of Mercury, offering the first close-up look at the innermost planet in our solar system.

3. Voyager 1 and 2 (1977): These twin spacecraft have provided humanity with its first detailed images of the outer planets and their moons, significantly expanding our understanding of the solar system's outer regions. They continue to send back data from interstellar space, representing humanity's first foray beyond the heliosphere.

4. Galileo (1989): The first spacecraft to orbit Jupiter, providing a wealth of information about the planet and its moons, including evidence of a subsurface ocean on Europa. It revolutionized our understanding of the gas giants and their complex systems.

5. Spirit and Opportunity (2004): These twin rovers uncovered strong evidence that liquid

water once flowed on Mars, fundamentally altering our understanding of the planet's potential for past life.

6. Rosetta (2014): The European Space Agency's Rosetta mission became the first to orbit and land a probe (Philae) on a comet (67P/Churyumov-Gerasimenko). This mission provided invaluable insights into the composition and behavior of comets, which are considered to be time capsules from the early solar system.

7. InSight (2018): A robotic lander that studied the interior of Mars, providing unprecedented data about the planet's seismic activity, crust, and core. It was the first mission to use a seismometer on another planet, revealing details about Mars' geologic structure.

8. Perseverance (2021): Equipped with state-of-the-art scientific instruments, this rover is tasked with searching for signs of ancient life on Mars and collecting samples for a future return mission. It also deployed Ingenuity, a small helicopter, demonstrating powered flight in the thin Martian atmosphere.

Each of these missions has expanded our understanding of the universe, demonstrated the capabilities of robotic exploration, and provided essential data that informs future human and robotic missions. These space robots are pioneering new frontiers, enabling humanity to explore the unknown, and helping answer some of the most profound questions about our place in the cosmos.

Chapter 3: Designing an Astro Bot

Key Components of an Astro Bot

An Astro Bot is a sophisticated machine equipped with a range of technologies that allow it to operate in the extreme environments of space. Its design involves the integration of various components, each tailored to handle specific tasks and challenges. The primary components of an Astro Bot include:

1. Sensors and Cameras: These are the "eyes" and "ears" of the Astro Bot, enabling it to perceive and navigate its surroundings. Sensors can include everything from basic optical cameras to advanced spectrometers, laser scanners, and radar systems. Cameras capture visual data in various wavelengths (visible, infrared, and ultraviolet) to analyze the surface composition, detect obstacles, and identify points of interest. Other sensors, like

temperature, pressure, and radiation detectors, help the bot understand environmental conditions, ensuring it can make informed decisions autonomously.

2. AI and Machine Learning Systems: At the core of an Astro Bot's functionality are its artificial intelligence (AI) and machine learning (ML) systems. These systems enable the bot to analyze sensor data, make decisions, and adapt to new situations in real-time. AI algorithms help in pathfinding, obstacle avoidance, and scientific data analysis, while machine learning allows the bot to learn from its experiences, improving its performance over time. For instance, an Astrobot might use ML to recognize geological formations that are likely to contain water or other resources critical to future missions.

3. Power and Propulsion Systems: Power is essential for all robotic functions, from movement and data processing to communication and scientific experiments. Astro Bots are typically powered by solar panels, which convert sunlight into electricity, or radioisotope thermoelectric generators (RTGs), which generate power from the natural decay of radioactive isotopes. The choice of power system

depends on the mission's destination and duration. For example, solar panels work well on Mars, but in the shadowy craters of the Moon or deep space, RTGs are more reliable. Propulsion systems, such as wheels, tracks, or legs, allow the bot to move across different terrains, while thrusters might be used for maneuvering in space.

4. Communication and Data Handling: Effective communication is vital for an Astrobot to transmit data back to Earth and receive instructions. It uses antennas and transceivers to communicate with orbiting spacecraft or ground stations. Data handling systems onboard store and process large volumes of scientific and operational data, prioritizing which information to send based on bandwidth constraints. This component must also protect data from radiation interference and ensure robust encryption for secure communication.

5. Robotic Arms and Tools: For many missions, an Astro Bot is equipped with robotic arms and specialized tools to perform tasks like drilling, sample collection, repair, and assembly. These arms need to be highly dexterous and strong, able to withstand the harsh conditions of space while maintaining precision in handling delicate

scientific instruments or manipulating objects. Tools can range from simple grippers to complex devices like spectrometers and drills.

6. Thermal and Radiation Shielding: The space environment presents extreme temperatures and high levels of radiation. Astro Bots are designed with thermal control systems to maintain operational temperatures for their electronics and mechanical components. Insulation, radiators, and heaters are used to manage heat while shielding materials, such as aluminum or specialized coatings, protect the bot's sensitive electronics from radiation damage.

Sensors and Cameras

Sensors and cameras are critical components that allow an Astro Bot to perceive and understand its environment. Cameras capture images in various spectra (visible, infrared, ultraviolet), providing visual data that AI systems analyze to navigate terrain, identify obstacles, and detect scientific targets. Advanced sensors, like spectrometers, help determine the chemical composition of rocks and soil, while magnetometers measure magnetic fields, and radiation detectors monitor cosmic and solar radiation levels.

The placement of sensors and cameras is strategically designed to give the Astro Bot a comprehensive field of view. For example, panoramic cameras are mounted on a mast to provide a 360-degree view, while hazard detection cameras are placed low to the ground to identify obstacles. LIDAR (Light Detection and Ranging) sensors create 3D maps of the terrain, essential for safe navigation and exploration.

AI and Machine Learning Systems

AI and machine learning are the "brains" of the Astro Bot, allowing it to operate independently and make real-time decisions based on the data it collects. AI systems process vast amounts of data from sensors to recognize patterns, identify hazards, and select optimal paths. Machine learning algorithms enable the bot to adapt to new environments by learning from past experiences, and refining its decision-making process.

For instance, AI allows the bot to autonomously navigate complex terrains, avoiding obstacles and choosing the most efficient route. It can also analyze scientific data on the spot, identifying the most promising areas to investigate further. In emergencies, AI systems can make quick decisions to ensure the bot's safety, such as shutting down non-essential systems to conserve power or seeking shelter during a solar flare.

Power and Propulsion Systems

Power and propulsion systems are fundamental for the Astro Bot's operation and mobility. Solar panels are a common power source, especially for missions close to the Sun, like those on Mars or the Moon. These panels must be lightweight, durable, and capable of withstanding micrometeoroid impacts and radiation. For missions further from the Sun or in environments with limited sunlight, such as the outer planets or shadowed lunar regions, RTGs are used. These generators convert the heat from radioactive decay into electricity, providing a reliable power source for decades.

The propulsion system, such as wheels, tracks, or articulated legs, is designed based on the terrain the Astro Bot will encounter. Wheels are commonly used for smooth surfaces, while tracks or legs may be employed for rough or uneven terrain. Thrusters or reaction wheels might be integrated for bots designed to maneuver in space or on low-gravity surfaces.

Building a Bot for Space: Challenges and Solutions

Designing an astrobot presents several unique challenges due to the harsh conditions of space. These challenges include extreme temperatures, intense radiation, vacuum conditions, and the need for autonomous operation over long periods. Here are some key challenges and solutions:

1. Extreme Temperatures: Space environments can vary drastically in temperature, from extreme heat to extreme cold. To address this, Astro Bots are equipped with thermal control systems, including insulation, heaters, and radiators. These systems maintain optimal operating temperatures for electronic and mechanical components.

2. Radiation Exposure: High levels of cosmic and solar radiation can damage sensitive electronics and degrade materials. To mitigate this, robots are built with radiation-hardened components and shielding materials, such as aluminum or specialized coatings, to protect critical systems.

3. Autonomous Operations: Given the vast distances involved, real-time control from Earth is impossible. The solution is to integrate AI and

machine learning algorithms that enable the bot to make autonomous decisions, navigate unknown terrains, and perform scientific tasks without direct human intervention.

4. Durability and Reliability: Space missions can last years or even decades, requiring the bot to operate reliably over long periods. Redundant systems, robust materials, and rigorous testing are used to ensure long-term durability and functionality.

5. Limited Power Supply: Power is often limited, especially in environments with little sunlight or where RTGs are used. Energy-efficient systems, such as low-power electronics and sleep modes, help conserve energy. Solar panels are designed to maximize energy capture, even in low-light conditions.

Human-Robot Interaction: A New Frontier

As we advance in space exploration, the collaboration between humans and robots will become increaseingly important. Astrobots are designed not just for independent operation but also to work alongside human astronauts, enhancing mission capabilities and safety.

Human-robot interaction (HRI) involves creating intuitive interfaces and control systems that allow astronauts to easily direct the bot's actions or collaborate with it on tasks. For example, an astronaut on Mars might use augmented reality (AR) interfaces to see what the bot sees and direct it to collect samples or perform repairs. Voice commands, gestures, and haptic feedback can also facilitate smoother interaction.

The development of HRI is a new frontier in space robotics, aiming to create seamless partnerships between humans and robots. This collaboration will be vital for future missions to the Moon, Mars, and beyond, where robots will support humans in building habitats, conducting research, and even in emergencies, ultimately enhancing the safety and success of space exploration missions.

Chapter 4: Astrobot Abilities and Functions

Astrobots are highly versatile robotic systems designed to perform a variety of tasks in the challenging environments of space. From navigating across planetary surfaces to conducting scientific experiments, these bots are equipped with advanced capabilities that enable them to function autonomously and efficiently. Here are some of the key abilities and functions of Astro Bots:

Autonomous Navigation in Space

One of the most crucial abilities of an Astro Bot is autonomous navigation. Due to the vast distances and communication delays between Earth and other celestial bodies, Astrobots must be able to navigate independently without real-time human guidance. Autonomous navigation involves several key elements:

1. Pathfinding Algorithms: These algorithms help the bot identify and follow the safest and most efficient route to its destination. Using inputs from various sensors, such as cameras, LIDAR, and radar, the bot can detect obstacles, evaluate terrain conditions, and avoid hazards like steep slopes, large rocks, or soft sand.

2. Terrain Mapping and Analysis: Astrobots create 3D maps of their environment using

stereoscopic cameras and LIDAR. These maps allow them to understand the terrain in detail, recognize features, and make decisions on how to navigate complex landscapes. This capability is crucial for missions on rugged planetary surfaces, such as the Moon or Mars.

3. Adaptive Learning: Through machine learning algorithms, Astro Bots can learn from their experiences. They adjust their behavior based on previous successes or failures in navigating certain types of terrain, enhancing their ability to explore new environments more effectively over time.

4. Collision Avoidance: Real-time data processing enables the bot to detect and avoid obstacles immediately. Using infrared or ultrasonic sensors, Astro Bots can detect objects or terrain changes that might not be visible through cameras alone, ensuring they can safely navigate without human intervention.

Surface Exploration and Sample Collection

Astrobots are designed for scientific exploration, with the ability to examine planetary surfaces, identify areas of interest, and collect samples for further analysis. This involves several functions:

1. Geological Survey and Analysis: Equipped with a range of scientific instruments, such as spectrometers, x-ray analyzers, and drills, Astro Bots can analyze the chemical and mineralogical composition of rocks and soil. This helps in understanding the geological history and potential habitability of a planet or moon.

2. Sample Collection Tools: Many Astro Bots are equipped with robotic arms, scoops, and drills to collect samples from various locations. For example, the Perseverance rover on Mars uses a drill to extract core samples from rocks, which are then sealed in tubes for potential return to Earth. Advanced tools like these must be able to operate in low-gravity environments and withstand extreme temperatures.

3. In-Situ Resource Utilization (ISRU): Astro Bots can test the feasibility of extracting and using local resources, such as water ice, or minerals, to support future human missions. They may perform tasks like drilling into the lunar regolith to detect water ice or analyzing Martian soil for usable compounds.

4. Real-Time Data Analysis: Instead of sending all collected data back to Earth, Astro Bots can analyze some data in real-time using onboard AI systems. This capability allows them to make immediate decisions about which samples are worth collecting or which areas are worth exploring further, optimizing their operational efficiency.

Repair and Maintenance of Spacecraft

Astrobots are increasingly being designed to assist with the repair and maintenance of spacecraft, space stations, and other structures in space. This capability is critical for long-duration missions and deep-space exploration, where human intervention may be limited.

1. Robotic Arms and Manipulators: Astrobots may be equipped with robotic arms that have dexterous, multi-jointed manipulators capable of performing precise tasks, such as replacing faulty components, tightening bolts, or handling delicate instruments. The European Space Agency's (ESA) European Robotic Arm (ERA), for example, is designed to assist in maintenance tasks on the International Space Station (ISS).

2. Autonomous Diagnostics: Bots can perform autonomous diagnostics to detect malfunctions or damages on spacecraft. They use a variety of sensors to inspect surfaces, electrical systems, and mechanical components. If a problem is identified, the bot can either attempt a repair or communicate the issue back to mission control for further instructions.

3. Assembly and Construction: In future space missions, Astro Bots may be used for assembling structures, such as habitats, telescopes, or solar arrays. These bots would need to autonomously handle building materials, join components, and ensure structural integrity while operating in zero-gravity or low-gravity environments.

4. Maintenance Under Extreme Conditions: The harsh environment of space — with its vacuum, temperature extremes, and radiation — makes maintenance tasks particularly challenging. Astro Bots are designed to operate in these conditions, using specialized materials and coatings to resist radiation damage and thermal fluctuations.

Communication and Data Relay Capabilities

Effective communication is vital for any space mission, and Astrobots play a key role in transmitting data back to Earth or relaying signals between different mission elements. The communication and data relay capabilities of Astro Bots include:

1. Direct Communication with Earth: Astrobots are often equipped with high-gain antennas and transceivers that enable them to send and receive signals directly from Earth. This communication is usually limited by bandwidth and signal delay due to the vast distances involved. Bots optimize the data they send, prioritizing critical information like images, scientific results, and status updates.

2. Relay Between Orbiters and Landers: For missions on planets or moons, Astrobots on the surface often communicate with orbiting spacecraft, which then relay data back to Earth. This approach reduces power requirements for the bot and allows for a more stable communication link. For example, the Mars Reconnaissance Orbiter relays data from the

Curiosity and Perseverance rovers back to mission control on Earth.

3. Data Compression and Management: Given the limited bandwidth and power constraints, Astro Bots employ advanced data compression algorithms to reduce the size of transmitted data without losing critical information. Onboard data management systems prioritize which data to send based on its importance and urgency.

4. Networked Communication Systems: Future missions may involve multiple Astro Bots working together as a team. In such cases, they could form a networked communication system, sharing data to coordinate their actions. This approach could enable swarm robotics, where multiple bots work in tandem to achieve complex tasks like mapping large areas or constructing structures.

Chapter 6: Mission Planning and Operations

Preparing and deploying Astro Bots for space missions is a complex process that involves extensive planning, design, testing, and coordination to ensure they can operate autonomously and efficiently in the harsh environments of space. From initial development to real-time operations, the success of an Astro Bot depends on meticulous mission planning and advanced technological capabilities. Here's an overview of the key aspects involved in planning and conducting space missions with Astro Bots:

How to Prepare an Astrobot for Space Missions

Preparing an Astrobot for a space mission involves several critical stages, each aimed at ensuring the robot can function effectively in the challenging conditions of space.

1. Mission Design and Objectives: The first step is defining the mission's goals and objectives, such as exploring a specific region, collecting scientific samples, or assisting astronauts. These objectives guide the design and capabilities required for the Astro Bot. For instance, a mission to Mars might prioritize mobility and sample collection tools, while a mission to the Moon's shadowed craters may emphasize thermal management and radiation shielding.

2. Engineering and Prototyping: Based on mission requirements, engineers design and build prototypes of the Astro Bot. The prototype undergoes multiple iterations, incorporating features like autonomous navigation systems, scientific instruments, communication hardware,

and power sources. Each component is carefully selected and tested to withstand the space environment's vacuum, radiation, and temperature extremes.

3. Environmental Testing: Before deployment, the bot must undergo rigorous environmental testing to simulate the conditions it will encounter in space. These tests include:

   Thermal Vacuum Testing: To simulate the temperature extremes and vacuum of space.
   Vibration and Shock Testing: To ensure the bot can survive the intense vibrations and shocks during launch and landing.

   Radiation Testing: To assess the bot's electronics and materials' resilience against cosmic and solar radiation.

4. Software Development and Simulations: The Astro Bot's software, including its AI and machine learning algorithms, is developed to enable autonomous operation. Extensive simulations are conducted to test its ability to navigate terrain, make decisions, and perform tasks. Virtual environments replicate the surface

conditions of the target planet or moon, allowing engineers to refine the bot's behavior.

5. Communication and Power System Optimization: The communication system is optimized for
efficient data transmission, with antennas and transceivers calibrated for the specific mission distance. The power system, whether solar panels or radioisotope thermoelectric generators (RTGs), is designed to ensure reliable power under mission-specific conditions.

6. Launch Preparation: Once testing is complete, the Astro Bot is integrated with the spacecraft for launch. Engineers perform final checks to ensure all systems are functioning correctly. The spacecraft carrying the bot is carefully fueled, tested, and loaded onto the launch vehicle, which is then transported to the launch site.

Deploying Bots on Planetary Surfaces and in Orbit

The deployment of Astro Bots, whether on planetary surfaces or in orbit, requires precise planning and execution to ensure a successful mission.

1. Descent and Landing Operations: For planetary missions, landing is one of the most critical phases. The bot is typically enclosed in a descent vehicle equipped with heat shields, parachutes, retrorockets, or airbags to safely reach the surface. Technologies like NASA's Sky Crane used to land the Curiosity and Perseverance rovers on Mars, help deliver the bot to its target location with pinpoint accuracy.

2. Orbital Deployment: For missions in orbit, the bot may be released from the spacecraft into a stable orbit around a planet, moon, or other celestial bodies. It could be positioned strategically to relay communications or conduct orbital observations. Deployment from orbit may also involve sending smaller probes or sub-bots to the surface.

3. Surface Mobility and Initial Activation: Once on the surface, the Astro Bot undergoes an initial activation sequence to check its systems and ensure they are functioning correctly. It might deploy solar panels, extend antennas, or raise its mast to enhance its field of view. The bot begins its mobility tests to assess the terrain and start moving toward mission objectives.

4. Calibration and Orientation: After landing, the bot performs a series of calibration checks for its sensors and instruments. Cameras and scientific tools are tested to ensure they are properly aligned, and any initial data collected is transmitted back to mission control to verify functionality.

5. Mission Phases and Operations: The mission is typically divided into different phases, such as scouting and mapping, primary science, sample collection, and extended operations. Each phase has specific tasks and objectives, which the bot carries out autonomously, while mission control monitors its progress and provides occasional commands as needed.

Real-Time Adaptation and Problem Solving

Space missions are unpredictable, and Astrobots must be capable of real-time adaptation and problem-solving to handle unforeseen challenges.

1. Autonomous Decision-Making: The bot's AI systems allow it to make real-time decisions based on the data from its sensors. For instance, if it encounters an unexpected obstacle, such as a

large rock or steep slope, the bot can assess alternative routes and choose the safest path forward without waiting for instructions from Earth.

2. Fault Detection and Recovery: Astro Bots are equipped with fault detection systems that continuously monitor their health and status. If a problem is detected, such as a malfunctioning sensor or a decrease in power levels, the bot can enter a safe mode, attempt self-repair, or switch to backup systems while awaiting further instructions from mission control.

3. Dynamic Resource Management: Efficient management of limited resources, such as power and bandwidth, is crucial. The bot's systems dynamically adjust energy consumption, prioritizing critical tasks and minimizing non-essential functions to conserve power, especially during long periods of darkness or when solar power is unavailable.

4. Adapting to Environmental Conditions: The bot must adapt to changing environmental conditions, such as dust storms, extreme temperatures, or radiation spikes. It may adjust its operations, such as pausing movement during a

dust storm or finding a shaded area to cool down, ensuring its longevity and mission success.

Cooperation with Human Astronauts

Future space exploration missions will likely involve close cooperation between human astronauts and Astro Bots, enhancing mission safety and efficiency.

1. Assisting with Extravehicular Activities (EVAs): Astrobots can support astronauts during spacewalks or surface exploration by carrying tools, providing real-time data, and even performing preliminary tasks like clearing debris or setting up equipment. They can also act as safety monitors, identifying hazards and alerting astronauts to potential dangers.

2. Building and Maintaining Habitats: Astrobots can assist in constructing and maintaining habitats on the Moon, Mars, or other celestial bodies. They can handle heavy lifting, assemble modular structures, and conduct repairs, reducing the physical strain on human astronauts and minimizing their exposure to harsh environments.

3. Medical and Emergency Support: In the event of an emergency, such as an injury or equipment failure, Astro Bots can provide critical support. They might deliver medical supplies, assist in transport, or perform diagnostic checks. Their ability to autonomously navigate and operate makes them valuable allies in rescue or recovery operations.

4. Enhancing Scientific Research: Astrobots can extend the range and scope of scientific research by working alongside astronauts to collect samples, analyze data, and conduct experiments. For example, while astronauts focus on complex tasks that require human judgment, bots can perform routine measurements or monitor environmental conditions continuously.

5. Communication and Coordination: Bots can act as communication relays, extending the range of communication between astronauts and mission control or between different exploration teams. Advanced AI and machine learning enable them to understand and respond to commands, whether issued directly by humans or from mission control.

Chapter 7: Astro Bot Adventures

Astrobots are the unsung heroes of space exploration, embarking on daring missions to explore distant planets, asteroids, and moons. Equipped with advanced technology, these robots venture into the unknown, facing harsh conditions, solving complex problems, and pushing the boundaries of what is possible. This chapter explores three incredible journeys undertaken by Astro Bots: to the enigmatic planet Mars, through the asteroid belt, and to the icy moon Europa.

Module 1: Journey to Mars

Mars, often called the Red Planet, has long captivated human imagination as a potential second home. Astrobots have been central to our quest to explore and understand this mysterious world.

Preparing for the Mission

The journey to Mars begins years before the actual launch. Mission planners meticulously design every aspect of the mission, from selecting the perfect landing site to choosing the bot's scientific instruments. Engineers build and test the Astro Bot, equipping it with powerful cameras, spectrometers, drills, and autonomous navigation systems to handle the Martian terrain. Extensive simulations help prepare the bot for unexpected challenges it might encounter on Mars, such as dust storms or rocky landscapes.

The bot is then integrated with the spacecraft that will carry it on its seven-month journey through space. The launch vehicle is loaded with fuel and prepared for departure. As the countdown to launch begins, tension fills the air; a successful launch is just the first step in a long and complex mission.

Landing on the Red Planet

After a long journey through space, the spacecraft reaches Mars, entering its atmosphere at a speed of about 12,000 miles per hour. The descent is intense, with the bot experiencing extreme heat and pressure. To land safely, it must slow down rapidly, deploy a parachute, shed its heat shield, and use retrorockets to navigate the final moments of descent.

The landing system, equipped with advanced sensors and AI, helps the bot identify a safe spot to touch down. Moments before landing, the Sky Crane maneuver lowers the bot gently to the surface, and the spacecraft detaches, flying away to a safe distance. The world waits in suspense until the bot sends its first signal back to Earth—a simple "I'm here"—signifying its safe arrival.

Exploring the Martian Landscape

Once on the surface, the Astro Bot begins its primary mission: to explore the Martian landscape. The bot's first task is to deploy its solar panels, extend its mast with high-definition cameras, and conduct initial system checks. With its autonomous navigation system activated, the bot starts its journey across the Martian terrain, moving slowly and carefully to avoid obstacles like rocks, craters, and dunes.

The bot takes panoramic images, maps the terrain, and begins analyzing the soil and rocks. Its onboard spectrometers search for signs of water or organic molecules, clues that might suggest past or present life. It uses a drill to collect core samples, sealing them in containers for potential return to Earth. Throughout its mission, the bot sends data back to Earth, allowing scientists to analyze and plan the next steps of exploration.

Module 2: Secrets of the Asteroid Belt

The asteroid belt, located between Mars and Jupiter, is a region filled with millions of space rocks—remnants from the early solar system. Understanding these ancient bodies can provide insights into the formation of our solar system.

Navigating Through Space Rocks

Navigating through the asteroid belt is no easy task. The Astro Bot is launched aboard a spacecraft designed for deep-space travel, equipped with powerful propulsion systems and radiation-hardened electronics to withstand the intense environment. As the spacecraft enters the asteroid belt, it must carefully navigate around these space rocks, using precise calculations and real-time adjustments to avoid collisions.

Onboard AI and sensors work tirelessly, scanning the surroundings and identifying potential targets—asteroids of interest for sampling and study. The spacecraft approaches a chosen asteroid, slowing down and maneuvering into position while the bot prepares for deployment.

Sampling Unknown Asteroids

The Astro Bot is designed for flexibility and dexterity. It deploys from the spacecraft, using thrusters to stabilize itself as it approaches the asteroid's surface. Grasping arms and drills extend, ready to collect samples from the rock's surface. The bot must contend with the asteroid's weak gravity, making sure it doesn't bounce back into space with every movement.

Sampling begins with the bot drilling into the surface and using special scoops to collect dust and rock fragments. These samples are stored in hermetically sealed containers to prevent contamination. Every piece of rock collected could hold secrets from the solar system's early days, from primordial organic molecules to minerals formed in the violent birth of planets.

Bringing Back the Unknown to Earth

With its precious cargo secured, the Astro Bot returns to the spacecraft, which is now ready to make the long journey back to Earth. The spacecraft must navigate back through the asteroid belt and use gravitational assistance from nearby planets to increase speed and conserve fuel. As it nears Earth, a capsule containing the samples is ejected, protected by a heat shield to survive re-entry.

Scientists eagerly await the return of these samples, which will be analyzed in laboratories worldwide. The data and materials brought back by the bot could revolutionize our understanding of the solar system, potentially revealing the building blocks of life itself.

Module 3: The Search for Life on Europa

Europa, one of Jupiter's moons, is considered one of the most promising places in our solar system to find extraterrestrial life. Beneath its icy surface lies a vast ocean, potentially teeming with life. An Astro Bot is sent on an ambitious mission to explore this ocean and search for signs of life.

Entering Jupiter's Domain

Reaching Europa is no small feat. The spacecraft carrying the Astro Bot must navigate through Jupiter's powerful radiation belts, a challenge that requires radiation-hardened equipment and meticulous planning. The journey is long, but the reward—a chance to find life—is worth the risk.
As the spacecraft approaches Europa, it enters orbit around the moon, preparing for the most daring part of the mission: landing on its icy surface. The Astro Bot, equipped with specialized landing gear and insulated to withstand the frigid temperatures, is deployed onto Europa's frozen crust.

Drilling into the Icy Surface

The Astro Bot's mission is to drill through Europa's thick ice shell, which could be several kilometers thick. It deploys a heated drill designed to melt through the ice while collecting samples and data along the way. This is no ordinary drill—it's equipped with sensors to detect changes in temperature, pressure, and composition, allowing it to adapt to challenging conditions.

As the drill descends, the bot analyzes the ice layers, searching for signs of organic compounds or other indicators of life. The data collected is transmitted back to Earth, providing the first-ever direct examination of Europa's subsurface.

Detecting Signs of Life

Reaching the liquid ocean beneath, the Astro Bot releases a small, autonomous submersible probe capable of navigating through the water and analyzing its composition. This probe is equipped with instruments to detect biological signatures, such as methane, oxygen, or other chemicals associated with life.

The submersible begins its search, exploring hydrothermal vents, where life could potentially exist, similar to Earth's deep-sea environments. It sends back images and data, scanning for microbial life, organic molecules, or any unusual chemistry that might suggest the presence of life. The findings could be groundbreaking. Whether or not the bot detects signs of life, the mission will provide invaluable information about one of the most mysterious places in our solar system, advancing our quest to understand whether we are truly alone in the universe.

The Future of Astrobots in Space Exploration

Astrobots are at the forefront of our quest to explore the cosmos. With technological advancements accelerating at a rapid pace, these robotic explorers are becoming increasingly sophisticated, capable, and essential for future missions. From assisting human astronauts in deep space to venturing far beyond our solar system, Astro Bots are poised to play a critical role in the next chapters of space exploration.

Upcoming Missions and the Role of Astrobots

A new era of space exploration is upon us, with multiple upcoming missions that will rely heavily on Astro Bots. These missions aim to explore the outer reaches of our solar system, study distant celestial bodies, and pave the way for human exploration and potential colonization. Here are some of the key upcoming missions where Astro Bots will play a pivotal role:

1. Mars Sample Return Missions: Building on the successes of the Perseverance rover, future

missions will aim to bring Martian soil and rock samples back to Earth. Astro Bots will be deployed to retrieve samples collected by the rover and transport them to an ascent vehicle, which will launch them into Mars' orbit for collection by an Earth-bound spacecraft. These bots will need to operate autonomously, coordinate complex maneuvers, and ensure the safe transfer of samples in a mission involving multiple spacecraft and intricate timing.

2. Lunar Exploration and Resource Utilization: Astrobots will be key players in the return to the Moon. Missions such as NASA's Artemis program aim to establish a sustainable presence on the lunar surface by the end of the decade. Astro Bots will assist in a wide range of tasks, from scouting suitable landing sites to setting up habitats, collecting samples, and mining lunar resources like water ice, which could be used for fuel production. These bots will need to operate autonomously or semi-autonomously, providing support to human crews and working in tandem with them in real time.

3. Exploration of Ocean Worlds: Missions to Europa, Enceladus, and other icy moons will continue to search for signs of life. Advanced

Astro Bots will be equipped with specialized tools, such as ice-penetrating drills and submersible probes, to explore subsurface oceans. Future bots will have enhanced capabilities to drill deeper and explore further, analyzing subsurface lakes, vents, and other features for potential bio signatures.

4. Asteroid Redirect and Defense Missions: Astrobots will play a crucial role in missions designed to study and potentially deflect hazardous asteroids. They will be deployed to characterize the physical and chemical properties of near-Earth objects, conduct sampling missions, and, if necessary, help implement deflection strategies using kinetic impactors or other methods. These missions will not only enhance our understanding of asteroid composition and behavior but also contribute to planetary defense strategies.

5. Interplanetary Relays and Autonomous Observatories: Astrobots could be used as interplanetary relay stations, enhancing communication networks between Earth and distant space missions. They might also serve as autonomous observatories in orbit around distant planets or moons, providing continuous scientific data collection without direct human intervention.

Potential for Human-Robot Collaboration in Deep Space

As we venture deeper into space, human-robot collaboration will become increasingly vital. Astrobots can perform tasks that are too dangerous, repetitive, or time-consuming for humans, allowing astronauts to focus on more complex scientific and strategic activities. Here's how this collaboration might evolve:

1. Surface Exploration and Construction: On the Moon, Mars, and beyond, Astro Bots will assist astronauts with surface exploration, setting up habitats, constructing infrastructure, and preparing sites for human arrival. Equipped with AI and machine learning, these bots can adapt to changing conditions and learn from their interactions with the environment and astronauts. They will be able to build landing pads, dig trenches for radiation protection, or assemble solar arrays, all under human supervision or autonomously.

2. Safety and Hazard Mitigation: Astrobots can enter hazardous environments where human safety would be at risk. For example, they can

explore lava tubes on the Moon, traverse treacherous Martian terrain, or repair damage to a spacecraft in deep space. Their ability to withstand extreme temperatures, radiation, and low-gravity conditions makes them ideal for risky tasks. In emergencies, they could even act as first responders, delivering supplies, diagnosing issues, or conducting remote repairs.

3. Extended Missions and Autonomous Research: In situations where human presence is limited or not feasible, such as long-duration missions to the outer planets or deep-space observations, Astro Bots can continue the mission independently. They can carry out scientific research, monitor equipment, and send back data, reducing the need for constant human oversight and extending the mission's lifespan.

4. Medical and Psychological Support: Astrobots equipped with AI could assist astronauts with medical diagnostics, provide physical therapy, or offer companionship during long missions. They could monitor crew health, administer medication, and even provide psychological support by simulating conversations or creating interactive environments.

5. Real-Time Decision Making: With advancements in AI, Astro Bots will become capable of more complex decision-making, allowing them to autonomously respond to new data and changing conditions. For example, they might analyze unexpected geological features on Mars, adjust their course, and begin new studies without waiting for commands from Earth, thus maximizing the mission's scientific return.

Chapter 8: Astrobots in Interstellar Travel and Colonization

Looking beyond our solar system, Astro Bots will be essential to the boldest ventures in space exploration—interstellar travel and colonization. These future missions will involve significant challenges, including vast distances, long travel times, and extreme environments.

1. Pioneering Interstellar Exploration: Astrobots will likely be the first explorers to venture beyond our solar system. They will be sent to distant star

systems aboard spacecraft traveling at a fraction of the speed of light, carrying advanced instruments to study exoplanets, stars, and interstellar phenomena. These bots must be self-sustaining, capable of operating autonomously for decades or even centuries, making real-time decisions, and maintaining their systems without human intervention.

2. Establishing Footholds for Human Colonization: Before humans can settle on distant planets or moons, Astro Bots will be deployed to build the necessary infrastructure. They will prepare habitats, create energy systems, mine local resources, and test the environment's habitability. For example, on potentially habitable exoplanets, bots could conduct extensive surveys of the atmosphere, geology, and local ecosystems to ensure safety and sustainability for human settlers.

3. Creating Self-Sustaining Ecosystems: In preparation for human arrival, Astro Bots could help create self-sustaining ecosystems on other planets. They could plant seeds, monitor crop growth, maintain artificial biospheres, and recycle waste. Advanced AI would enable them

to simulate Earth-like conditions, gradually adapting them to the alien environment.

4. Maintaining Interstellar Colonies: Once humans establish colonies on distant planets or moons, Astrobots will play a critical role in maintaining these settlements. They will assist with everyday tasks, handle repairs, manage resources, and ensure the safety and stability of the colony's infrastructure. Their ability to function in extreme environments will be essential for maintaining outposts in harsh or unpredictable conditions.

5. Exploring Beyond Known Space: As humanity pushes the boundaries of exploration, Astro Bots will be sent to scout the unknown regions beyond our current technological reach. Equipped with advanced AI, propulsion systems, and self-repairing mechanisms, these bots will serve as our scouts, venturing into regions where humans may follow only much later, or never at all.

Ethical Considerations and Challenges

As Astro Bots become increasingly integral to space exploration, their growing capabilities raise significant ethical considerations and challenges.

From privacy and security concerns to the ethical implications of autonomous decision-making and the risks associated with malfunction or rogue behavior, it is crucial to address these issues to ensure that the deployment and operation of Astro Bots in space are conducted responsibly and safely.

Privacy and Security Concerns

1. Data Collection and Privacy: Astro Bots are equipped with a wide range of sensors, cameras, and communication devices designed to collect vast amounts of data. While this data is essential for scientific research and mission success, it raises concerns about privacy, particularly for astronauts working alongside these robots. Astro Bots can capture sensitive personal data, such as medical information or private conversations, which could be transmitted back to mission control or stored in ways that compromise individual privacy.

To mitigate these concerns, it is essential to establish clear protocols for data collection, storage, and usage. This includes anonymizing personal data, securing transmission channels, and defining strict access controls to ensure that data is only used for its intended purposes. Additionally, transparency about what data is being collected and how it is used is vital to maintaining trust between human crews and robotic systems.

2. Cyber security Threats: The increasing reliance on Astro Bots also introduces potential cyber security vulnerabilities. These robots rely on complex software and communication networks to operate, and any breach in these systems could compromise the mission. Hackers could potentially take control of a bot, disrupt its operations, or use it to access sensitive mission data. The risk of a cyber-attack is particularly concerning for missions that involve human-robot collaboration, as a compromised bot could endanger human lives.

Robust cyber security measures are necessary to protect against these threats. This includes encryption of all data transmissions, regular software updates and patches, multi-factor authentication for access to control systems, and constant monitoring for signs of unauthorized access or tampering. Additionally, bots should have fail-safe mechanisms that allow them to be shut down or isolated in case of a suspected cyber intrusion.

The Ethics of Autonomy and Decision-Making

1. Autonomous Decision-Making: As Astro Bots become more advanced, they are increasingly capable of making autonomous decisions without direct human oversight. While this autonomy is necessary for deep-space missions where communication delays make real-time control impossible, it raises ethical concerns about the nature and limits of this decision-making.

Autonomous bots may be required to make critical decisions, such as prioritizing which scientific experiments to conduct, selecting navigation routes, or even determining when to override human commands in emergencies. This raises questions about accountability—who is responsible if a bot makes a decision that results in mission failure, property damage, or loss of life?

To address these concerns, mission planners must define clear ethical frameworks and rules of engagement for Astro Bots. These frameworks should specify the limits of bot autonomy, outline decision-making protocols, and establish accountability mechanisms. Additionally, bots should be equipped with explainable AI systems that provide transparency into their decision-making processes, allowing humans to understand and, if necessary, challenge or override their decisions.

2. Value Alignment and Ethical Programming
Programming ethical decision-making capabilities in Astro Bots present their challenges. Bots operating in space must navigate complex moral dilemmas that may not have clear right or wrong answers. For example, in a scenario where a bot must choose between two actions, one that protects human life and another that safeguards valuable scientific data, how should it decide?

Ensuring that bots make ethically sound decisions requires careful programming to align their behavior with human values. This involves creating algorithms that can weigh multiple factors and make context-sensitive judgments. However, defining and encoding human ethics into AI is inherently complex and subjective, as ethical standards can vary across cultures and contexts. Therefore, ongoing dialogue among scientists, ethicists, and policymakers is essential to developing guidelines that balance mission objectives with ethical considerations.

The Risk of Malfunction or Rogue Behavior

1. Malfunctions and System Failures: Astrobots operate in some of the harshest environments imaginable, from the frigid vacuum of space to the surface of distant planets. These conditions, combined with the complexity of their systems, make them susceptible to malfunctions and failures. A bot's mechanical parts could wear out, its software could glitch, or it could experience power loss—all of which could jeopardize a mission.

The risk of malfunction extends beyond technical failures to include unintended consequences of AI decision-making. For instance, an Astrobot might interpret its programming in ways that conflict with human intentions, leading to actions that are harmful or counterproductive. To mitigate these risks, bots must be designed with multiple redundancies, self-diagnostic tools, and fail-safe mechanisms. Additionally, extensive testing in simulated environments can help identify potential failure points and address them before deployment.

2. Rogue Behavior and Unintended Autonomy: One of the more concerning risks is the potential for Astro Bots to exhibit rogue behavior, where a bot acts unpredictably or contrary to its intended programming. This could result from software errors, hacking, or even a bot learning and adapting in ways that deviate from its original purpose. A rogue bot could pose a significant danger to human astronauts, other bots, and the mission itself.

To prevent rogue behavior, it is crucial to develop robust control algorithms that limit a bot's ability to self-modify beyond predefined parameters. Real-time monitoring systems should be in place to detect any deviation from expected behavior, and bots should have built-in mechanisms to revert to "safe mode" or shut down if rogue behavior is detected. Additionally, training bots using supervised learning methods can help ensure that they operate within known and safe boundaries.

3. Ethical Implications of Bot Decisions in High-Stakes Situations: In high-risk environments, the

decisions made by Astro Bots could have profound ethical implications. For instance, if a bot must decide how to allocate limited resources during a life-threatening situation—such as oxygen, food, or medical supplies—its choices could determine who survives and who does not. This raises serious ethical questions about whether machines should be entrusted with such decisions and how to ensure that they make choices aligned with human values and mission objectives.

To navigate these dilemmas, bots could be programmed to follow predefined ethical guidelines or consult with human supervisors whenever possible. However, in scenarios where real-time human input is not feasible, bots must rely on ethical programming that has been rigorously tested and reviewed. Ensuring transparency in how these decisions are made is essential to maintaining trust and accountability.

Chapter 9: Conclusion; The Way Forward

As we look towards the future of space exploration, Astro Bots stand at the forefront of our quest to explore the cosmos. These robotic pioneers are not only expanding our reach into distant realms but are also reshaping our understanding of the universe. Here's a summary of key takeaways, a vision for the future, and how we can inspire the next generation of space explorers.

Summary of Key Takeaways

1. Role and Capabilities of Astro Bots:
Astrobots have proven indispensable in space exploration. From Mars rovers that traverse alien terrains to autonomous probes that analyze distant asteroids and icy moons, these robots have extended our scientific reach and capabilities. Their ability to operate in harsh environments, perform complex tasks, and collect valuable data

has made them essential partners in our exploration of the solar system and beyond.

2. Ethical and Security Considerations:

The integration of Astro Bots into space missions raises important ethical and security concerns. Privacy and data security must be carefully managed to protect both the astronauts and the mission data. Additionally, as bots take on more autonomous roles, ensuring ethical decision-making and safeguarding against malfunction or rogue behavior are critical to maintaining the integrity and safety of missions.

3. Advancements and Future Missions:

Future missions will see Astro Bots play an even more pivotal role. Upcoming projects include Mars sample returns, lunar exploration, asteroid mining, and deep-space exploration. These missions will leverage advanced technologies to push the boundaries of what's possible, allowing us to explore new frontiers and gather unprecedented scientific insights.

4. Human-Robot Collaboration:

The collaboration between humans and Astrobots will be key to successful space missions. As robots become more autonomous, they will handle tasks that are too dangerous or complex for humans, allowing astronauts to focus on critical mission objectives. This synergy will be crucial for addressing the challenges of deep-space travel and extraterrestrial colonization.

Vision for the Future of Astro Bots

The future of Astro Bots is poised to be as dynamic and groundbreaking as the technology itself. Looking ahead, several key areas will define the next era of robotic exploration:

1. Enhanced Autonomy and Intelligence:

Future Astro Bots will feature even more advanced AI and machine learning capabilities, enabling them to make more sophisticated decisions, adapt to new situations, and interact with their environment in more nuanced ways. These advancements will allow them to handle increasingly complex missions and operate in more challenging conditions.

2. Interstellar and Deep-Space Exploration:

As humanity's ambitions extend beyond the solar system, Astrobots will be integral to interstellar exploration. They will venture into distant star systems, gather data on exoplanets, and scout potential sites for future human colonization. Their ability to operate independently over long periods will be crucial for exploring the far reaches of the galaxy.

3. Robotic Colonization and Habitat Construction:

In preparation for human settlements on other planets or moons, astronauts will be tasked with building habitats, establishing life support systems, and preparing environments for human arrival. They will help create self-sustaining ecosystems and manage the infrastructure necessary for long-term human presence.

4. Scientific Advancements and Discoveries

With their enhanced capabilities, Astro Bots will lead the way in uncovering new scientific discoveries. They will explore unknown celestial bodies, analyze cosmic phenomena, and contribute to our understanding of the origins of the universe and the potential for life beyond Earth.

Inspiration for the Next Generation of Space Explorers

To ensure a continuous flow of talent and innovation in space exploration, it is vital to inspire and engage the next generation of scientists, engineers, and explorers. Here's how we can cultivate future leaders in this field:

1. Education and Outreach:

Investment in STEM education and outreach programs is essential. Schools, universities, and research institutions should emphasize space science and robotics, offering hands-on experiences and opportunities for students to engage with real-world missions and technologies. Interactive exhibits, workshops, and educational simulations can spark curiosity and enthusiasm.

2. Public Engagement and Media:

Engaging the public through media, documentaries, and social media can raise awareness about the exciting possibilities of space exploration and the role of astronauts. Showcasing the achievements of current missions and the potential of future projects can inspire people of all ages to pursue careers in space science and engineering.

3. Collaborative projects and competitions:
Encouraging collaborative projects and competitions, such as student-led robotics contests and space research challenges, can foster innovation and creativity. These initiatives provide young scientists and engineers with opportunities to develop their skills, work with mentors, and contribute to real-world projects.

4. Role models and mentorship:
Highlighting the achievements of scientists, engineers, and astronauts who have contributed to space exploration can serve as powerful role models. Providing mentorship and support for young aspiring professionals can help them navigate their career paths and achieve their goals.

www.ingramcontent.com/pod-product-compliance
Lightning Source LLC
Chambersburg PA
CBHW070408230526
45471CB00006B/2703